Elmer E. Rasmuson Library Cataloging in Publication Data

Bowman, Timothy Dale
Field guide to bird nests and eggs of Alaska's coastal tundra / Timothy D. Bowman. – Fairbanks, Alaska : Alaska Sea Grant College Program, University of Alaska Fairbanks, 2004.

   p. : ill. ; cm. - (Alaska Sea Grant College Program ; SG-ED-44)

   NOAA National Sea Grant NA16RG2321, A/161-01.

1. Birds—Nests—Alaska. 2. Birds—Eggs—Alaska. I. Title. II. Series: Alaska Sea Grant College Program ; SG-ED-44.

QL675.B69 2004

ISBN 1-56612-085-3

Credits

This book was designed and produced using funds from the U.S. Fish and Wildlife Service, Alaska region. It is published by the Alaska Sea Grant College Program, supported by the U.S. Department of Commerce, NOAA National Sea Grant Office, grant NA16RG2321, A/161-01; and by the University of Alaska Fairbanks with state funds. The University of Alaska is an affirmative action/equal opportunity institution. Alaska Sea Grant publication SG-ED-44.

Sea Grant is a unique partnership with public and private sectors combining research, education, and technology transfer for public service. This national network of universities meets changing environmental and economic needs of people in our coastal, ocean, and Great Lakes regions.

ALASKA SEA GRANT COLLEGE PROGRAM
UNIVERSITY OF ALASKA FAIRBANKS
P.O. Box 755040 • 205 O'Neill Bldg.
Fairbanks, Alaska 99775-5040
Toll free (888) 789-0090
(907) 474-6707 • FAX (907) 474-6285
http://www.uaf.edu/seagrant/

U.S. FISH AND WILDLIFE SERVICE
DIVISION OF MIGRATORY BIRD MANAGEMENT
1011 East Tudor Road
Anchorage, Alaska 99503-6199
(907) 786-3443 • Fax (907) 786-3641

Field Guide to Bird Nests and Eggs of Alaska's Coastal Tundra

Timothy D. Bowman
U.S. Fish and Wildlife Service
Anchorage, Alaska
Published by Alaska Sea Grant College Program

Table of Contents

Introduction 2
Photo Credits 7

LOONS
Yellow-billed Loon 8
Common Loon 9
Red-throated Loon 10
Pacific Loon 11

WATERFOWL and CRANE
Tundra Swan 12
Sandhill Crane 13

GEESE
Parting Shots—Geese by Tail Patterns 14
Quick Reference for Dark Goose Nests 15
Taverner's Canada Goose 16
Snow Goose 17
Emperor Goose 18
Greater White-fronted Goose 19
Cackling Canada Goose 20
Brant 21

EIDERS
Common Eider 22
Spectacled Eider 23
King Eider 24
Steller's Eider 25

DUCKS
Black Scoter 26
White-winged Scoter 27
Canvasback 28
Red-breasted Merganser 29
Greater Scaup 30
Long-tailed Duck (Oldsquaw) 31
Northern Shoveler 32
Mallard 33
Northern Pintail 34
Green-winged Teal 35
American Wigeon 36

JAEGERS, GULLS, and TERNS
Pomarine Jaeger 37
Parasitic Jaeger 38
Long-tailed Jaeger 39
Glaucous Gull 40
Glaucous-winged Gull 41
Mew Gull 42
Sabine's Gull 43
Aleutian Tern 44
Arctic Tern 45

OWLS
Snowy Owl 46
Short-eared Owl 47

PTARMIGAN
Willow Ptarmigan 48
Rock Ptarmigan 49

ALCIDS
Black Guillemot 50

SHOREBIRDS
Whimbrel 51
Bar-tailed Godwit 52
Marbled Godwit 53
Wilson's Snipe 54
Long-billed Dowitcher 55
American Golden-Plover 56
Pacific Golden-Plover 57
Black-bellied Plover 58
White-rumped Sandpiper 59
Ruddy Turnstone 60
Black Turnstone 61
Pectoral Sandpiper 62
Buff-breasted Sandpiper 63
Stilt Sandpiper 64
Baird's Sandpiper 65
Rock Sandpiper 66
Dunlin 67
Western Sandpiper 68
Semipalmated Sandpiper 69
Least Sandpiper 70
Semipalmated Plover 71
Red Phalarope 72
Red-necked Phalarope 73

PASSERINES
Redpolls 74
Snow Bunting 75
Lapland Longspur 76
Savannah Sparrow 77
Golden-crowned Sparrow 78
Yellow Wagtail 79

Sizing Chart for Unpatterned Eggs 80
Sizing Chart for Patterned Eggs 81

Field Guide to Bird Nests and Eggs of Alaska's Coastal Tundra

INTRODUCTION

The purpose of this guide is to allow rapid and accurate identification of nests encountered during biological investigations and monitoring studies of coastal tundra areas of Alaska. Its intended audience is field biologists and naturalists, although others may find it useful. The need for this field guide became apparent after several years of training and supervising crews of biologists on the Yukon-Kuskokwim Delta. This regionally habitat-focused guide is lightweight, weatherproof, and simple to use in the field, and is intended to supplement but not replace other existing field guides to birds, eggs, and nests.

The guide is sequenced (more or less) from large birds (large eggs) to small birds (small eggs), unlike most guides that present birds in taxonomic order. The rationale for this unconventional presentation is to reduce the time required to identify a nest in the field. Many people, including biologists, have not memorized the taxonomic order, but they can surely compare relative sizes of eggs. The egg sizing charts at the back of this guide are intended to quickly reduce the number of possible species to which an egg belongs based on egg length and general appearance and reduce the chance of overlooking a species. Whenever possible, similar species are placed on opposing pages to help comparisons.

Included for each of the 71 species are representative photos of adult birds usually of the parent most likely to incubate eggs, typical nesting habitat, nests, eggs (at actual size), and, for most waterfowl, breast feathers and wings showing diagnostic plumage patterns. Egg color, pattern, and markings can be quite variable within a species, so do not expect a perfect match. Also, because the nest photos were taken under different lighting conditions, the color representation is variable. For many species, photos of more than one egg are provided to illustrate variation. This variation may be enough that eggs of one species are almost indistinguishable from those of a related species.

Variation in egg shape and size (brant)

Variation in egg color and markings (willow ptarmigan)

Duck and goose nests are typically lined with many down feathers and a few breast feathers. Patterns and coloration of down and breast feathers can be diagnostic for some species although, like eggs, considerable variation exists. It is necessary to look at several feathers within each nest to better ensure correct identification.

Despite the tips and reference material provided here, there may be some nests that prove difficult to identify. It is possible that the species was not included in the guide because it was considered too rare in these habitats to warrant inclusion (e.g., surfbird, sanderling, yellow warbler). Sometimes nest identification can be particularly challenging if a nest has been destroyed or if more than one species has contributed eggs or feathers to the nest. By collecting samples of nest materials (breast feathers and down, egg fragments if present), biologists may be able to determine species identification by later comparison with a reference collection of materials. If the nest contains eggs, measure at least two eggs using calipers or sizing charts on pages 80-81 of this book. Or take photos.

The surest way to positively identify a nest to species is to recognize the adult bird that flushes from the nest. Among waterfowl, owls, ptarmigan, and most passerines (songbirds), only the female incubates eggs and the male is often absent from the nest area. For most shorebirds and gulls, either sex may incubate eggs. For ducks, it is useful to note wing plumage patterns, particularly the speculum on the trailing part of the inner wing. The hen's flight pattern as it flushes from the nest may be helpful to distinguish a dabbler, which jumps from the nest at a high angle, from a diver, which skitters across the water or flies at a low angle.

Dabbling ducks jump off water or nests

Diving ducks usually skitter on takeoff

The geographic scope of this guide is limited to the coastal tundra regions of Alaska from the Alaska Peninsula north along the west coast and east across the Arctic Coastal Plain. This includes low-lying moist tundra, as well as associated dwarf-shrub and upland tundra habitats. Although these areas make up a small portion of Alaska, they encompass a huge coastal strip, longer than the coastline of all the Pacific coastal states and Baja California combined and similar in land area to the state of Washington. These Alaska areas include some of the most productive bird nesting areas in North America, including the only North American or worldwide breeding areas for several species. This guide also includes a few species that more commonly nest in coastal alpine tundra, and species that nest in more inland habitats that are easily confused with one or more coastal-nesting species.

Relative abundance is given for four general regions: Southwest Alaska, including the north side of the Alaska Peninsula and Bristol Bay area; Yukon-Kuskokwim Delta; Northwest Alaska, including the Seward Peninsula, Kotzebue Sound, and Selawik lowland areas; and the Arctic Coastal Plain.

Dotted line represents approximate inland extent of habitats covered by this guide.

Relative abundance refers to the likelihood of encountering nesting birds of that species in coastal tundra habitats. Because nest detectability varies among species due to nest characteristics and parental behavior, the likelihood of encountering nesting birds may not necessarily reflect the likelihood of encountering nests. For example, a species may be a common nester but because its nests are so well hidden, few may be found. Thus, relative abundance categories are defined as follows:

Abundant–high density of nesting birds in a large portion of the region.
Common–nesting birds likely to be found at moderate density throughout most of the region.
Uncommon–nests at low density or localized only.
Rare–nesting birds rarely encountered.

Average egg sizes are based on measurements of eggs from Alaska when available, otherwise western Canada, or published sources including *The Birds of North America* series (The Birds of North America, Inc., Philadelphia, PA); P.J. Baicich and C.J.O Harrison, 1997, *A Guide to the Nests, Eggs and Nestlings of North American Birds*; and C.A. Reed, 1965, *North American Bird Eggs*. Maximum/minimum egg length represents a 90% confidence interval (90% of eggs are within the specified size range). Typical clutch sizes are reported here; observed clutches may be smaller or larger due to predation, dump nesting (more than one female lays eggs in a nest), and whether or not the bird is still laying eggs when its nest is found.

CAUTION AND PRUDENCE

Birds are particularly sensitive to disturbance and vulnerable to predation while nesting. If you do not have a legitimate reason to disturb a nest, then please leave it alone. To reduce impact on nesting birds, biologists should leave the nest and surrounding vegetation undisturbed whenever possible and leave the nest area quickly. Mammalian predators such as foxes may follow scent trails to nests, and predatory birds such as gulls and jaegers will take eggs and young from exposed nests. With the exception of waterfowl, all birds leave their eggs uncovered when they vacate the nest. Waterfowl often cover their eggs with nest material. If you must move vegetation concealing the nest or eggs, or if you encounter nests that have been exposed when the adult flushed, you can reduce the chance of a predator detecting the nest by sprinkling dry vegetation over the nest and eggs. For waterfowl nests, you can tease apart and fold over the perimeter of the nest lining to cover the eggs while the female is off the nest.

The collection of eggs of some species for subsistence is currently legal for residents in some areas covered by this guide (see http://alaska.fws.gov/ambcc). Otherwise, eggs, nests, and feathers of migratory birds may be collected only with a permit from the U.S. Fish and Wildlife Service. The purpose of this book is to furnish a reference that can be used as an aid in the study of nesting birds, and it is not the intent or desire of the author that this book should stimulate illegal or indiscriminate egg collection. Admire and learn from what nature has to offer, but please tread lightly so that future generations can enjoy these birds.

ACKNOWLEDGMENTS
All photographs were taken by Tim Bowman unless otherwise noted. Thanks to the dozens of biologists/photographers who graciously contributed photos to this guide (see credits page). Photographs should not be reproduced without permission from the photographer. Thanks to the University of Puget Sound's Slater Museum of Natural History for allowing the author to photograph their egg collection and for shorebird wing photos. Rene Corado provided measurements of eggs from the Western Foundation of Vertebrate Zoology. Chris Hitchcock assisted with design. Financial support for production of the guide was provided by the U.S. Fish and Wildlife Service (Migratory Bird Management Division and Yukon Delta National Wildlife Refuge). Special thanks to Chris Dau, Dan Gibson, Bob Gill, Sue Keller, Rick Lanctot, Brian McCaffery, Randy Meyers, Bob Stehn, and Declan Troy for reviewing drafts of all or portions of this guide.

AUTHOR BIOGRAPHY
Tim Bowman has been involved in ornithological investigations throughout the United States and Canada since 1983. He came to Alaska in 1989, employed by the U.S. Fish and Wildlife Service, to study the effects of the *Exxon Valdez* oil spill on bald eagles. Since 1993, he has been involved primarily in studies of waterfowl. His studies have included the molting ecology of black ducks in Labrador, studies of gull predation on waterfowl on the Yukon-Kuskokwim Delta, and studies of sea duck biology and migration in the Pacific. He coordinated the nest survey used to monitor reproductive effort and population trends of birds on the Yukon-Kuskokwim Delta, and has trained observers for nine years. His dedication to improving the reliability and accuracy of scientific data, and interest in photography, inspired him to develop this field guide to bird nests and eggs of Alaska's coastal tundra.

He currently works for the U.S. Fish and Wildlife Service in Anchorage, Alaska, and is the U.S. Coordinator for the Sea Duck Joint Venture, a conservation partnership designed to improve our understanding of sea ducks and the reasons for declines in many of their populations.

PHOTO CREDITS

Initials	Name
BA	Brad Andres
BG	Bob Gill
BJ	Brian Johnson
BJM	Brian J. McCaffery
BM	Bob Montgomery
CE	Craig Ely
CH	Chuck Hunt
CLG	Cheri L. Gratto-Trevor
CM	Catherine Moiteret
CPD	Chris P. Dau
CRE	Charles R. Eldermire
DES	David E. Saffine
DHW	David H. Ward
DKM	Dennis K. Marks
DLS	Dave L. Scobie
DM	Dave Menke
DN	Dave Nysewander
DP	Dennis Paulson
DR	Dan Roby
DRR	Dan R. Ruthrauff
DT	Declan Troy
DW	© Doug Wechsler/VIREO
EH	Eric Hopson
EHM	Edward H. Miller
FT	© F. Truslow/VIREO
FWS	U.S. Fish and Wildlife Service
GD	George Divoky
GK	Gary Kramer
GMB	Guillaume M. Bouteloup
GWB	Gerry W. Beyersbergen
HC	© H. Cruickshank/VIREO
HCK	© H.C. Kyllingstad/VIREO
HLD	H. Loney Dickson
HM	Heather Moore
JBF	Julian B. Fischer

Initials	Name
JJ	Jim Johnson
JL	Joe Liebezet
JPP	John P. Pearce
JRC	Jesse R. Conklin
JSF	Jeff S. Fair
JTP	Jeff T. Pelayo
JTT	John T. Toppenberg
JW	Jeff Wasley
JWP	John W. Prather
KF	Krista Fahy
KN	Kristina Norstrom
LQ	Lori Quakenbush
LT	Lee Tibbetts
MF	Michael Forsberg
MJP	Mike J. Petrula
MM	Mark Masteller
MN	Mark Nyhof
MP	Matt Perry
MS	Marnie Shepherd
NAB	Naomi A. Bargmann
OWJ	O. Walter Johnson
PDM	Philip D. Martin
PM	Pam Miller
RDT	Roger D. Titman
RL	Rick Lanctot
RWC	R. Wayne Campbell
RWO	Ron W. Opsahl
SJL	© S.J. Lang/VIREO
SS	Susan Savage
TD	Tammy Dixon
TJM	Tim J. Moser
TKM	Tamara K. Mills
TM	Tohru Mano
TS	Ted Swem
TV	© T. Vezo/VIREO

Yellow-billed Loon
(Gavia adamsii)

Large loon with a yellow bill. Nests on bare tundra, usually on a raised site at water's edge; sometimes on a small island or mound of vegetation. Nest is shallow scrape with little or no vegetative lining and usually damp. Eggs brown or yellowish-olive with many dark brown spots and blotches. Incubation by both sexes.

Relative Abundance: Southwest–none, Y-K Delta–none, Northwest–rare, Arctic Coastal Plain–rare, but locally more frequent in parts of National Petroleum Reserve and Colville River Delta.

Typical Clutch: 2 eggs
Average Egg Size: 91 x 55.2 mm

Actual Size

Common Loon
(Gavia immer)

Large loon with "checkerboard" back, dark bill, and all dark head. Nest is a mound of aquatic vegetation, wet grasses, and mosses that averages about 55 cm (22 in) in diameter. Usually within a meter of water, sometimes concealed, often on island or amassed atop vegetation in shallow water. Eggs dark greenish-brown with black spots. Incubation by both sexes.

Relative Abundance: Southwest–common, Y-K Delta–rare, Northwest–none, Arctic Coastal Plain–none, but occurs farther inland in foothills on north side of Brooks Range.

Typical Clutch: 2 eggs
Average Egg Size: 89.5 x 57.4 mm

Max
Min

Actual Size

LOONS

Red-throated Loon
(Gavia stellata)

Loons flush at long distances when approached, so be alert as you approach ponds. Red-throated loon has a raspy, rapid "*quacky*" call. Nests and eggs of red-throated and Pacific loons are virtually identical. Red-throated loon eggs are, on average, slightly smaller than those of Pacific loons, and red-throated loons tend to nest on smaller water bodies. However, neither of these traits is diagnostic as there is much variation. Nests are built on land at water's edge or over water atop a heap of vegetation. No down or feathers in nests. Incubation by both sexes.

Relative Abundance: Southwest–uncommon, Y-K Delta–uncommon, Northwest–uncommon, Arctic Coastal Plain–uncommon

Typical Clutch: 2 eggs
Average Egg Size: 73.9 x 45 mm

Actual Size

Pacific Loon
(Gavia pacifica)

Loons flush at long distances when approached, so be alert as you approach ponds. Pacific loon has a mournful, wailing call and a high pitched sharp "*whiip*" just before diving, as well as a guttural growl when disturbed. Nests and eggs of red-throated and Pacific loons are virtually identical and their sizes overlap. Nests are built on land at water's edge or over water atop a heap of vegetation. No down or feathers in nests. Incubation by both sexes.

Relative Abundance: Southwest–uncommon, Y-K Delta–common, Northwest–common, North Slope–common

Typical Clutch: 2 eggs
Average Egg Size: 76.2 x 47.2 mm

Max
Min

Actual Size

Tundra Swan
(Cygnus columbianus)

Adult usually leaves nest at long distances when approached. Nests in open grass-sedge meadows, uplands, and islands. Nest is a huge mound of vegetation with a hollow center. Nest usually lacks down, but may contain a few white breast feathers. No other eggs in this region are as large.

Relative Abundance: Southwest–common, Y-K Delta–common, Northwest–common, Arctic Coastal Plain–uncommon

Typical Clutch: 5 eggs
Average Egg Size: 107.3 x 68 mm

Actual Size

WATERFOWL

Sandhill Crane
(Grus canadensis)

Adults usually leave nest well in advance of your approach, but often stay in vicinity and are quite vocal. Nests in a variety of open habitats. Nest is usually a broad platform with a thin grass lining; diameter of 0.5-1 meter (1.5-3 ft). Nest is well camouflaged and easily overlooked. Eggs variable, but usually pale brown to light olive with irregular brown, reddish-brown, or gray markings. Incubation by both sexes.

Relative Abundance: Southwest–common, Y-K Delta–common, Northwest–common, Arctic Coastal Plain–rare

Typical Clutch: 2 eggs
Average Egg Size: 90.7 x 57.4 mm

Actual Size

WATERFOWL

Parting Shots...

In addition to head and body plumage, tail patterns can help distinguish among goose species.

Canada Goose
U-shaped white sub-terminal band

White-fronted Goose
Two white bands; terminal band is narrower

Brant
Single V-shaped white band

Emperor Goose
Wide single white terminal band

Quick Reference for Identification of Dark Goose Nests

START IN THIS ORDER	Brant	Canada	Emperor	White-fronted
DOWN COLOR	Darker gray	Darker gray	Lighter gray	Lighter gray
NEST MATERIAL	Down largely free of vegetation	Down mixed with grass	Down mixed with grass	Down mixed with grass
DOWN ABUNDANCE	Abundant	Abundant	Sparse	Sparse
BREAST FEATHERS	Mostly all gray; sometimes all white	Mostly all gray; sometimes all white	Light bluish-gray with rust-colored tip	White; distal half may be all black or mottled gray or black

WATERFOWL

Taverner's Canada Goose
(Branta canadensis taverneri)

Usually nests close to water, particularly on islands and peninsulas, or shorelines along rivers. Also nests on cliffs along rivers, sometimes in raptor stick nests. Nests often contain a large amount of down (but less than brant) with grass uniformly woven among down (unlike brant). Breast feathers are mostly uniformly gray with gray central vein, or occasionally all white.

Relative Abundance: Southwest–rare, Y-K Delta–rare coastally but common inland, Northwest–common, Arctic Coastal Plain–uncommon

Typical Clutch: 4-6 eggs
Average Egg Size: 78.4 x 53 mm

Actual Size

Snow Goose
(Chen caerulescens)

The only white goose you'll see in Alaska. Nests in open grassy areas by freshwater lakes or ponds, usually in small colonies. Nest is a hollow on ground lined with grasses or other plant material, white contour feathers and pale gray down. Eggs white or creamy-white.

Relative Abundance: Nests only on Arctic Coastal Plain—uncommon overall, but common locally (colonies)

Typical Clutch: 4-6 eggs
Average Egg Size: 79.7 x 52.8 mm

Actual Size

WATERFOWL

Emperor Goose
(Chen canagica)

Nests most commonly along pond shorelines and slough banks, sometimes farther from water in grass-sedge. Nests contain down mixed with grass; down is generally less abundant and lighter gray than that of brant and cacklers. Breast feathers have a light bluish-gray background with a rust-colored band near the tip of the feather and sometimes a narrow subterminal black band. The central shaft of the breast feather is usually white, except for distal third, which is gray. Eggs noticeably larger than those of brant and cacklers, but similar to white-fronted. Dump nesting is common and clutch sizes are highly variable.

Relative Abundance: Southwest–none, Y-K Delta–common, Northwest–rare, Arctic Coastal Plain–none

Typical Clutch: 4-6 eggs
Average Egg Size: 80.4 x 52.3 mm

Actual Size

Greater White-fronted Goose
(Anser albifrons)

More likely to nest farther from water than other geese; nest sites include dry sedge meadows, slough banks, uplands, and lake shores. Nests generally contain less down than brant and cacklers, but a similar amount to emperors. Down color is light gray but usually darker than emperors. Breast feathers are variable; may be entirely white or have a white background with an all black or black- or gray-mottled distal half (these feathers taken together form the "speckles" on the belly). Central shaft of breast feather is white nearly its entire length. Eggs noticeably larger than those of brant and cacklers, but similar to emperors.

Relative Abundance: Southwest–uncommon, Y-K Delta–abundant, Northwest–common, Arctic Coastal Plain–common

Typical Clutch: 4-7 eggs
Average Egg Size: 80.7 x 53.9 mm

Max

Min

Actual Size

WATERFOWL

Cackling Canada Goose
(Branta canadensis minima)

Usually nests close to water, particularly on islands and peninsulas. Nests often contain a large amount of down (but less than brant) with vegetation uniformly woven among down (unlike brant). Breast feathers are mostly gray, sometimes with variable amounts of white mixed with gray in distal half; occasionally all white. Central shaft of breast feather is white at base and gray in distal two-thirds.

Relative Abundance: Nests on Y-K Delta only—abundant

Typical Clutch: 4-6 eggs
Average Egg Size: 73.7 x 49.6 mm

Max
Min

Actual Size

Brant
(Branta bernicla)

Nests in colonies and as isolated pairs. Typically nests on islands and shorelines. Nests contain the largest amount of down of the four dark goose species. Down is largely free of vegetation and "crackles and snaps" when pulled apart. Breast feathers are usually light gray at base with gradually darker gray toward distal portion; more uniform in appearance than breast feathers of Canada geese. Central shaft of breast feather is white in proximal half and gray in distal half.

Relative Abundance: Southwest–none, Y-K Delta–abundant, Northwest–uncommon, Arctic Coastal Plain–uncommon (nests mostly on barrier islands)

Typical Clutch: 3-5 eggs
Average Egg Size: 71.1 x 47.9 mm

Max
Min

Actual Size

Common Eider
(Somateria mollissima)

Nests commonly along shoreline or on islands and peninsulas. Hen skitters away from nest. Compared to spectacled eider, body is larger, bill is heavier and longer, and wingbeat is slower. Wings have slight white bars on speculum. Hen will often circle back near the nest, sometimes with male—be alert. Down is mixed with grass and lighter gray than for spectacled eider. Breast feathers are similar to spectacled eider, but the barring is often more pronounced in common eider. Greenish eggs. Female often defecates on eggs when flushed.

Relative Abundance: Southwest–uncommon, Y-K Delta–uncommon, Northwest–common, Arctic Coastal Plain–uncommon (locally common on barrier islands)

Typical Clutch: 4-7 eggs
Average Egg Size: 75 x 49.8 mm

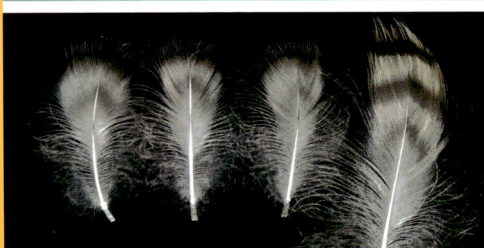

Actual Size

Spectacled Eider
(Somateria fischeri)

Nests commonly along shoreline or on islands and peninsulas. Hen flushes at low angle or skitters across water. Look for light eye patch; hen will often circle back near the nest, sometimes with male—be alert. Body and bill are smaller than common eider and wingbeat is more rapid. Down is dark gray, darker than that of common eider, and is mixed with grass. Contour feathers have a variable amount of barring. Greenish eggs. Female often defecates on eggs when flushed.

Relative Abundance: Southwest–none, Y-K Delta–uncommon, Northwest–rare, Arctic Coastal Plain–uncommon

Typical Clutch: 5-7 eggs
Average Egg Size: 67.7 x 45.1 mm

Actual Size

King Eider
(Somateria spectabilis)

Hen is slightly smaller than common eider and its bill is smaller. The bill is not feathered at base like spectacled eider. Usually nests in sedges along shorelines or on islands. Down is sooty brown with indistinct pale centers. Breast feathers largely unpatterned, light brown becoming darker brown toward tip. Greenish eggs.

Relative Abundance: Southwest–none, Y-K Delta–rare, Northwest–none, Arctic Coastal Plain–uncommon

Typical Clutch: 4-7 eggs
Average Egg Size: 67 x 44.6 mm

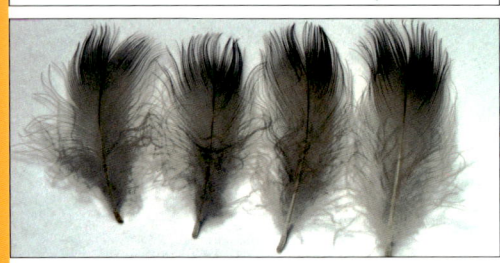

Actual Size

Steller's Eider
(Polysticta stelleri)

Nests at grassy margin of lakes and ponds. Hen flushes at low angle. Hen will often circle back near the nest, sometimes with male—be alert. Down is dark gray, almost black, mixed with grass. Usually deposits a few breast feathers, which are an unpatterned light brown becoming darker brown toward tip. Greenish eggs.

Relative Abundance: Southwest–none, Y-K Delta–rare, Northwest–none, Arctic Coastal Plain–rare (found primarily around Barrow)

Typical Clutch: 6-8 eggs
Average Egg Size: 59.1 x 41.4 mm

Actual Size

Black Scoter
(Melanitta nigra)

Nests more commonly near inland or upland lakes. Nests may be close to or far from water, usually in brushy transitions between dry lichen tundra and lake basins or drainages. Down tufts are dark brown with pale centers. Breast feathers are largely unpatterned, light brown at base grading into dark brown at the tips. Nests later than most other waterfowl, thus it may have an incomplete clutch if found during typical nest surveys.

Relative Abundance: Southwest–uncommon, Y-K Delta–uncommon, Northwest–uncommon, Arctic Coastal Plain–rare

Typical Clutch: 6-9 eggs
Average Egg Size: 67.3 x 46.5 mm

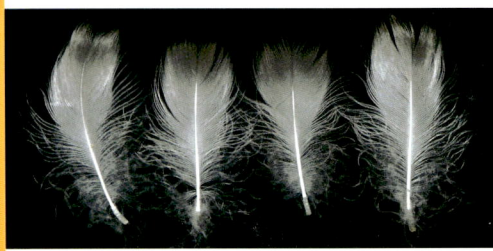

Actual Size

White-winged Scoter
(Melanitta fusca)

Typically nests more inland in forested or brushy habitats; rarely on coastal tundra. Body and wings are uniformly dark brown; has white speculum on wing. Female has sloping forehead and faint patches on cheek and behind bill. Nests at variable distances from water in dense brushy cover; nest often concealed by overhead vegetation or tree branches. Nest is a hollow lined with vegetation and down. Down tufts are dark brown with indistinct pale centers. Breast feathers (including central shaft) white at base becoming progressively dark brown-black in distal half. Eggs are creamy-buff.

Relative Abundance: Southwest–rare, Y-K Delta–none, Northwest–none, Arctic Coastal Plain–none

Typical Clutch: 7-11 eggs
Average Egg Size: 68.0 x 46.0 mm

Max

Min

Actual Size

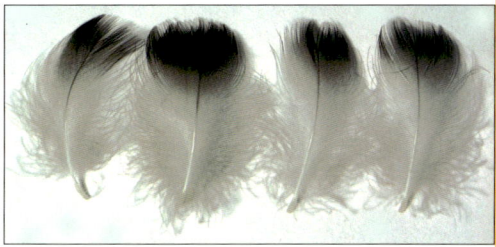

Canvasback
(Aythya valisineria)

Large-bodied diving duck with a wedge-shaped head profile. Usually builds nest over water in emergent vegetation. Nest bowl is deep; lined with abundant down mixed with dry leaves and grasses. Down tufts are light grayish-brown with indistinct pale centers and tips. Breast feathers are light grayish-brown tipped with white or tan and may exhibit fine gray speckling; central shaft is white in proximal third and gray distally; compare to greater scaup.

Relative Abundance: Southwest–none, Y-K Delta–rare, Northwest–rare, Arctic Coastal Plain–none

Typical Clutch: 7-9 eggs
Average Egg Size: 62.8 x 44.5 mm

Actual Size

Red-breasted Merganser
(Mergus serrator)

Body is mostly gray with an orange-brown head and ragged crest. Thin sawbill. Wing has a white speculum. Skitters across water when flushed from nest. Nests close to water in thickets or in cavities among rocks, roots, or riverbanks. Nest is a shallow depression lined with plant material and down, usually with overhead cover. Down tufts are dark gray, brown-tinged, with pale centers and tips. Breast feathers are mostly white; proximal portion of the central shaft is dark. Eggs creamy to greenish-buff.

Relative Abundance: Rare in all areas

Typical Clutch: 7-12 eggs
Average Egg Size: 63.4 x 44.6 mm

Max
Min

Actual Size

WATERFOWL

Greater Scaup
(Aythya marila)

Hen usually skitters across water when flushed. Wing shows speculum with white band extending into primaries. Nest is usually a deep bowl within 1 m (3 ft) of water, often concealed by overhead vegetation. Breast feathers are unevenly gray with white terminal band; central shaft is gray its entire length. Down is usually abundant, uniformly sooty-brown.

Relative Abundance: Southwest–common, Y-K Delta–common, Northwest–common, Arctic Coastal Plain–uncommon

Typical Clutch: 4-7 eggs
Average Egg Size: 63.5 x 41.9 mm

Max
Min

Actual Size

Long-tailed Duck
(Clangula hyemalis)

Previously named oldsquaw. If flushed, will fly low along water/land; no distinct markings on wings. Nests most often along shorelines, and on islands and peninsulas; sometimes farther from shore in short sedges. Nest is a deep bowl with abundant down mixed with little if any vegetation. Down tufts are dark brown, almost black, with pale centers. Breast feathers are mostly white, but are usually absent. Eggs yellowish or with faint olive or green tint.

Relative Abundance: Southwest–rare, Y-K Delta–uncommon, Northwest–common, Arctic Coastal Plain–common

Typical Clutch: 5-9 eggs
Average Egg Size: 54.4 x 38.2 mm

Actual Size

WATERFOWL

Northern Shoveler
(Anas clypeata)

Female looks much like a mallard, but bill is longer and wider. Wing speculum has white only on leading edge, and forewing appears pale. Nests in a variety of habitats, often in dense grass or sedge. Breast feathers are very diagnostic, with a large central dark spot surrounded by light area; central shaft is white except dark near distal tip. Down tufts are brown with light centers. Eggs usually creamy-buff to olive-tinted.

Relative Abundance: Southwest–uncommon, Y-K Delta–uncommon, Northwest–uncommon, Arctic Coastal Plain–rare

Typical Clutch: 8-12 eggs
Average Egg Size: 52.4 x 37.5 mm

Max
Min

Actual Size

Mallard
(Anas platyrhynchos)

Large dabbling duck with the ubiquitous nasal "*quack*." Nests in a variety of habitats, often in dense grass or sedge. Down tufts are brown with pale centers. Breast feathers are larger than those of most dabbling ducks, and have a brown band that extends to the tip of the breast feather; centers often have a cross pattern or alternating bars—can appear similar to pintail. Proximal half of central shaft is light, distal half is dark. Eggs usually pale green.

Relative Abundance: Southwest–uncommon, Y-K Delta–uncommon, Northwest–uncommon, Arctic Coastal Plain–rare

Typical Clutch: 9-11 eggs
Average Egg Size: 57.3 x 41.3 mm

Max
Min

Actual Size

WATERFOWL

Northern Pintail
(Anas acuta)

Note long slender neck and green in speculum. Nests in a variety of habitats, often in grass-sedge meadows or shorelines/slough banks; nest may be far from water. If hen is not seen, then breast feathers are the best diagnostic—note the wide brown stripe that narrows and extends to the tip; the exact pattern varies; central shaft is white except dark near distal tip. Down tufts are brown with inconspicuous white centers. Eggs variably yellowish-green.

Relative Abundance: Abundant in Southwest, Y-K Delta, and Northwest; common on Arctic Coastal Plain

Typical Clutch: 7-9 eggs
Average Egg Size: 54.3 x 38 mm

Actual Size

Green-winged Teal
(Anas crecca)

Tiny dabbling duck with tiny eggs. Most commonly nests in grass meadows and along shorelines and slough banks. If hen is not seen, then the small egg size and breast feathers are the best diagnostics. Breast feathers typically have a brown center spot with a pale interior. Down tufts are relatively small, dark beige, with diffuse white centers.

Relative Abundance: Southwest–uncommon, Y-K Delta–common, Northwest–common, Arctic Coastal Plain–rare

Typical Clutch: 9-12 eggs
Average Egg Size: 46 x 33.2 mm

Actual Size

WATERFOWL

35

American Wigeon
(Anas americana)

Medium-sized dabbling duck with a fairly short, bluish bill. Typically nests in grasses, usually open sites. Nest is a hollow lined with dry grass. Down is abundant; tufts are very dark, almost black, with conspicuously frosted tips. Breast feathers are mostly tan with light-colored tips. Cream-colored eggs.

Relative Abundance: Southwest–uncommon, Y-K Delta–uncommon, Northwest–common, Arctic Coastal Plain–rare

Typical Clutch: 8-11 eggs
Average Egg Size: 53.5 x 37.8 mm

Actual Size

Pomarine Jaeger
(Stercorarius pomarinus)

Both light and dark morphs exist. Central tail feathers are long, rounded, and twisted 90 degrees. Usually aggressive toward intruders. Bicolored bill. Nests on open flat tundra. Nest is a shallow scrape or hollow scantily lined with vegetation. Eggs buff, olive, or brown with spots and flecks of blackish brown. Incubation by both sexes.

Relative Abundance: Southwest–none, Y-K Delta–rare, Northwest–rare, Arctic Coastal Plain–rare, but regular around Barrow. Within its range it may be more common in years of rodent abundance.

Typical Clutch: 2 eggs
Average Egg Size: 62 x 44 mm

Max
Min

Actual Size

JAEGERS – GULLS – TERNS

Parasitic Jaeger
(Stercorarius parasiticus)

Plumage is variable. Tail streamers are pointed, but shorter than in long-tailed jaeger. Usually aggressive toward intruders. Nests on barren or dwarf-shrub tundra. Nest is a shallow hollow scantily lined with grass, lichen, or moss. Eggs usually olive or greenish with spots and flecks of blackish-brown; more heavily marked at larger end. Incubation by both sexes.

Relative Abundance: Uncommon in all areas

GWB

BA

Typical Clutch: 2 eggs
Average Egg Size: 57 x 40.5 mm

Max
Min

Actual Size

Long-tailed Jaeger
(Stercorarius longicaudus)

Tail streamers are long and pointed. Usually aggressive toward intruders. Nests on barren or dwarf-shrub tundra; nest is a shallow hollow scantily lined with lichen or moss. Eggs olive-green to olive-brown with spots, blotches, and scrawls; more heavily marked at larger end. Incubation by both sexes.

Relative Abundance: Southwest–rare, Y-K Delta–common, Northwest–common, Arctic Coastal Plain–common

Typical Clutch: 2 eggs
Average Egg Size: 54 x 38 mm

Max
Min

Actual Size

Glaucous Gull
(Larus hyperboreus)

Large gull with a heavy bill and no black tips on its wings. Eye has yellow iris. Most nests are built on islands, with lesser numbers on shorelines and peninsulas. Also nests on sandy barrier islands off coast of Arctic Coastal Plain. Nest is usually a volcano-shaped mound of vegetation, sometimes quite large. No down or feathers in nest. Eggs variably marked with spots, small blotches, speckling, or thin scrawls in dark olive-brown and blackish-olive, and pale markings in gray or violet. Incubation by both sexes.

Relative Abundance: Southwest–rare, Y-K Delta–common, Northwest–common, Arctic Coastal Plain–uncommon

Typical Clutch: 2-3 eggs
Average Egg Size: 75.6 x 51.9 mm

Actual Size

Glaucous-winged Gull
(Larus glaucescens)

Similar to glaucous gull, but rare from Y-K Delta northward, where glaucous gulls predominate. Distinguishable from glaucous gull by darker gray on wings and back and gray subterminal spots in wing tips. Eye has dark brown iris. Small numbers of glaucous-winged hybrids occur on Y-K Delta. Nests and eggs very similar to glaucous gull. Incubation by both sexes.

Relative Abundance: Southwest–abundant, Y-K Delta–rare, Northwest–none, Arctic Coastal Plain–none

Typical Clutch: 2-3 eggs
Average Egg Size: 73 x 51 mm

Max
Min

Actual Size

NAB

NAB

JBF

GMB

JAEGERS – GULLS – TERNS

Mew Gull
(Larus canus)

Medium-sized gull with black and white wingtips. Usually builds nest on islands, sometimes on peninsulas or shorelines of ponds. Nest is a shallow hollow built of vegetation. No down or feathers used in nest construction. May also nest in abandoned glaucous gull nests. Eggs variably marked with spots, blotches, specks, or short scrawls in brown, brownish-black, black, or fainter gray; more heavily marked at larger end. Incubation by both sexes.

Relative Abundance: Southwest–common, Y-K Delta–abundant, Northwest–common, Arctic Coastal Plain–none

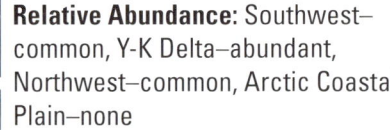

Typical Clutch: 2-3 eggs
Average Egg Size: 57.8 x 42.3 mm

Actual Size

Sabine's Gull
(Xema sabini)

Small tern-like gull with an all-dark head and striking black-white-gray wing pattern. Highly defensive and sometimes aggressive to intruders at nest sites. Prefers to nest on islands, peninsulas, and shorelines of ponds; sometimes nests in colonies. Nest is a scrape or shallow hollow lined with plant debris. No down or feathers used in nest construction. Eggs deep olive to buffish-olive, indistinctly and variably marked with blotches and scrawlings; often more heavily marked at larger end. Eggs can be confused with those of Arctic tern.

BG

Relative Abundance: Southwest–uncommon, Y-K Delta–common, Northwest–common, Arctic Coastal Plain–uncommon

MM

Typical Clutch: 2-3 eggs
Average Egg Size: 45.9 x 32.8 mm

Max
Min

Actual Size

Aleutian Tern
(Sterna aleutica)

Distinguished from Arctic tern by the white forehead and black bill and legs. Body is also darker than Arctic's. Seldom aggressive toward intruders, unlike Arctic tern. Call is a series of squeaky whistles. Usually nests on islands, peninsulas, and shorelines of ponds, often in colonies. Nest is typically a shallow depression in moss or other matted plant material. Eggs usually pale greenish or olive, boldly marked with blotches, spots, and specks (edges of markings are better defined than for similar eggs of Sabine's gull). Incubation by both sexes.

Relative Abundance: Southwest–common, Y-K Delta–rare, Northwest–uncommon, Arctic Coastal Plain–none

Typical Clutch: 2 eggs
Average Egg Size: 43.4 x 29.5 mm

Actual Size

Arctic Tern
(Sterna paradisaea)

Has black cap and nape, deep red bill, and long forked tail. Adults are often aggressive toward intruders near their nests. Usually nests on islands, shorelines of ponds, and peninsulas. Nest is typically a shallow scrape sparsely lined with plant material. Eggs usually pale greenish or olive, variably marked with blotches, spots, and specks (markings better defined than for similar eggs of Sabine's gull). Incubation by both sexes.

Relative Abundance: Southwest–common, Y-K Delta–common, Northwest–common, Arctic Coastal Plain–uncommon

Typical Clutch: 2 eggs
Average Egg Size: 40.8 x 29.7 mm

Max
Min

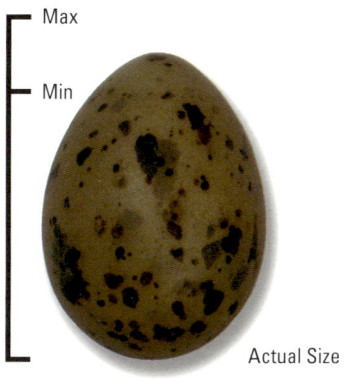

Actual Size

JAEGERS – GULLS – TERNS

Snowy Owl
(Bubo scandiacus)

FEMALE

MALE

Large, white owl with brown barring, more extensive in female. Nest is often on a hummock or rise overlooking tundra. Adult is often aggressive toward intruders. Nest is a scrape with little if any lining. Eggs are short-elliptical and white, sometimes becoming stained as incubation progresses. Incubation by female only. Young hatch asynchronously and therefore may be vastly different sizes.

Relative Abundance: Southwest–none, Y-K Delta–rare, Northwest–rare, Arctic Coastal Plain–rare. May be more common locally or in years when rodents are abundant.

Typical Clutch: 4-10 eggs
Average Egg Size: 58 x 45.6 mm

Actual Size

Short-eared Owl
(Asio flammeus)

Stiff wingbeats and wavering flight pattern. Nests in grass-sedge meadows. Nest is a shallow hollow sparsely lined with grass. Eggs white, smooth, and short-elliptical. Incubation by female only. Young hatch asynchronously and therefore may be vastly different sizes.

Relative Abundance: Rare in all areas. Cyclically more common in years when rodents are abundant.

Typical Clutch: 4-8 eggs
Average Egg Size: 39.8 x 32.2 mm

Max
Min

Actual Size

Willow Ptarmigan
(*Lagopus lagopus*)

Nests in a variety of habitats including grass-sedge meadows and uplands. Nest sometimes partially concealed by shrubby vegetation like dwarf birch. Eggs usually yellowish, heavily and irregularly blotched and mottled; quite variable. No down in nest, but may rarely have breast feathers. Incubation by female only.

Relative Abundance: Common in Southwest, Y-K Delta, and Northwest; uncommon on Arctic Coastal Plain

Typical Clutch: 6-11 eggs
Average Egg Size: 42.3 x 33.1 mm

Actual Size

Rock Ptarmigan
(Lagopus mutus)

MALE

FEMALE

Female looks like willow ptarmigan, but has slightly smaller bill. Typically nests on dry tundra. Nest is a hollow scrape, usually fully exposed, scantily lined with moss, lichen, and maybe a few feathers. No down in nest, rarely contains breast feathers. Eggs smooth and glossy; usually whitish or pale creamy-yellow irregularly blotched with reddish-brown. Incubation by female only.

Relative Abundance: Southwest–rare, Y-K Delta–rare, Northwest–uncommon, Arctic Coastal Plain–uncommon

Typical Clutch: 5-10 eggs
Average Egg Size: 42.5 x 30.9 mm

- Max
- Min

Actual Size

Black Guillemot
(Cepphus grylle)

Medium-sized alcid with black body and white wing patch. Nests in a cavity or among man-made debris or driftwood on beach; often colonial. Nest is an unlined hollow, sometimes surrounded by debris. Eggs buff or bluish-green, variably marked with spots and blotches of black, gray, or reddish-brown. Incubation by both sexes.

Relative Abundance: Nests only in isolated locations along Arctic Coastal Plain, primarily on some barrier islands or spits.

Typical Clutch: 2 eggs
Average Egg Size: 58.5 x 39.6 mm

Actual Size

Whimbrel
(Numenius phaeopus)

Large grayish-brown shorebird with long, distinctly downcurved bill. Call is a rapid and liquid "*quiquiquiqui*" with no change in pitch. Nests in tussocky or hummocky tundra, usually more inland than coastal. Nest is a shallow hollow, sparsely lined with plant material or lichen. Eggs pale green or olive or buff; marked with blotches and spots of brown or blackish-brown, more concentrated at larger end. Incubation by both sexes.

Relative Abundance: Southwest–rare, Y-K Delta–rare, Northwest–uncommon, Arctic Coastal Plain–rare

Typical Clutch: 4 eggs
Average Egg Size: 58.1 x 40 mm

Max
Min

Actual Size

SHOREBIRDS

Bar-tailed Godwit
(*Limosa lapponica*)

Large shorebird with long bill that curves slightly upward. Male has a deep rufous-cinnamon or chestnut coloration on underparts and head; female is grayer and larger than male. Call is a sharp "*kirrick*" or "*godwit*." Nest is a shallow hollow on sedge-dwarf shrub tundra, lined with grass. Eggs variably brown, with variable spots or blotchy markings usually concentrated at larger end. Incubation by both sexes.

Relative Abundance: Southwest–none, Y-K Delta–common, Northwest–uncommon, Arctic Coastal Plain–rare (may be more common locally)

Typical Clutch: 4 eggs
Average Egg Size: 54 x 38 mm

Actual Size

Marbled Godwit
(Limosa fedoa)

DW

Large shorebird, heavily barred and speckled, with long slightly upcurved bicolored bill. Nests on moss hummocks in sedge-bog areas. Nest is a shallow cup lined with horsetail, sphagnum moss, sedges, or leaves. Eggs pale buff or olive sparsely spotted and with small blotches and brown or purplish scrawls. Incubation by both sexes.

Relative Abundance: Nests only in Southwest–uncommon

HC

HC

CLG

Typical Clutch: 4 eggs
Average Egg Size: 58 x 40 mm

Max
Min

Actual Size

Wilson's Snipe
(Gallinago delicata)

Formerly common snipe. Stocky, long-billed, and short-winged with bold stripes down its back. Takeoff is explosive; flight rapid and zigzagging; usually calls with a harsh "*scaap*" when flushed. Nests in a variety of habitats from moist lowlands to upland tundra. Nest is a shallow hollow lined with grass; may have a canopy. Eggs pale green to olive; blotched, spotted, and speckled in brown and shades of gray. Incubation by female only.

Relative Abundance: Southwest–common, Y-K Delta–common, Northwest–common, Arctic Coastal Plain–rare

Typical Clutch: 4 eggs
Average Egg Size: 39.5 x 28 mm

Actual Size

Long-billed Dowitcher
(Limnodromus scolopaceus)

Straight sturdy bill twice the length of its head. In flight, white triangular patch above its tail. Nests on grass-sedge tundra and wet meadows. Nest is a well concealed, fairly deep cup lined with sedge, grasses, or small leaves. Eggs pale green or olive; speckled, spotted, and blotched with brown and purplish gray. Incubation by both sexes. Eggs, nests, and adults of **short-billed dowitcher** *(Limnodromus griseus)* are nearly identical to long-billed; in short-billed the wing feathers extend beyond tail unlike long-billed; short-billed's call is a melodic "*tu tu tu*" versus a single high-pitched "*keek*" for long-billed.

Relative Abundance: Uncommon in all areas; nests found in Southwest more likely to be short-billed dowitcher.

Typical Clutch: 4 eggs
Average Egg Size: 42.5 x 30 mm

Max
Min

Actual Size

SHOREBIRDS

55

American Golden-Plover
(*Pluvialis dominica*)

Dark feathers extend up nape and crown, unlike black-bellied plover with pale crown. Pacific golden-plover looks very similar, but the white on its neck extends farther down flanks, and it has white under-tail coverts. Alarm call of American golden-plover is a two syllable "*kleeep*" or "*klee-yeep*," the second note higher in pitch. Nests on upland tundra. Nest is a shallow scrape on ground, sparsely lined with grass or lichen. Eggs creamy-buff, heavily spotted and blotched in black. Incubation by both sexes.

Relative Abundance: Southwest–none, Y-K Delta–uncommon, Northwest–common, Arctic Coastal Plain–common

Typical Clutch: 4 eggs
Average Egg Size: 48 x 33 mm

Actual Size

Pacific Golden-Plover
(Pluvialis fulva)

Dark feathers extend up nape and crown, unlike black-bellied plover with pale crown. American golden-plover appears similar, but on the Pacific the white extends farther down flanks and it has white under-tail coverts. Alarm call is a one syllable "*peee*." Nests on upland tundra. Nest is a shallow scrape on ground, sparsely lined with grass or lichen. Eggs creamy-buff, heavily spotted and blotched in black. Incubation by both sexes.

Relative Abundance: Southwest–uncommon, Y-K Delta–rare, Northwest–rare, Arctic Coastal Plain–none

Typical Clutch: 4 eggs
Average Egg Size: 48 x 33 mm

Actual Size

SHOREBIRDS

57

Black-bellied Plover
(Pluvialis squatarola)

Large plover with white forehead and crown, unlike American or Pacific golden-plovers. Call is a mournful "*tlee-oo-ee.*" Nests in higher, dry tundra or upland areas. Nest is a shallow hollow on tundra, sparsely lined with lichen or moss. Eggs usually pale buff or grayish-stone, variably spotted and blotched in black; markings tend to be well defined and concentrated at larger end. Incubation by both sexes.

Relative Abundance: Southwest–rare, Y-K Delta–common, Northwest–rare, Arctic Coastal Plain–uncommon

Typical Clutch: 4 eggs
Average Egg Size: 52.8 x 36.6 mm

Max
Min

Actual Size

White-rumped Sandpiper
(Calidris fuscicollis)

Small shorebird with short straight bill, finely streaked gray breast; wing tips extend well past tip of tail. In flight, has white bar across base of tail. Nests in wet grass-sedge or mossy tundra, sometimes in a tussock or atop a hummock. Nest is a shallow hollow lined with dry grass or other vegetation. Eggs pale to olive green, spotted and blotched with reddish-brown, concentrated at larger end. Incubation by female only.

Relative Abundance: Nests only on Arctic Coastal Plain–uncommon

Typical Clutch: 4 eggs
Average Egg Size: 33.4 x 23.8 mm

Actual Size

SHOREBIRDS

Ruddy Turnstone
(Arenaria interpres)

Black and white head bib; red-orange legs. Bold, contrasting pattern on wings in flight. Call is a low pitched gutteral rattle. Nests in association with heath, often near water. Nest is a shallow hollow. Eggs pale green, bluish-green or light olive, irregularly blotched, spotted, streaked, and speckled with brown and paler gray markings. Markings are generally small and concentrated at larger end. Incubation by both sexes.

Relative Abundance: Southwest–none, Y-K Delta–uncommon, Northwest–uncommon, Arctic Coastal Plain–uncommon

Typical Clutch: 4 eggs
Average Egg Size: 40 x 29 mm

Actual Size

Black Turnstone
(Arenaria melanocephala)

Bold, contrasting pattern on wings in flight. Calls are a trilling "*skirr*" and a gutteral rattle, higher pitched than ruddy's. Incubation by both sexes. Nests usually within several meters of water. Nest is a cup-like hollow in grasses or sedges, lined with grass and often with a partial canopy of vegetation. Eggs usually olive or pale green; marked with spots, irregular blotches, and scrawls of olive-brown or blackish-brown.

Relative Abundance: Southwest–uncommon, Y-K Delta–abundant, Northwest–rare, Arctic Coastal Plain–none

Typical Clutch: 4 eggs
Average Egg Size: 41.1 x 28.9 mm

Max
Min

Actual Size

SHOREBIRDS

Pectoral Sandpiper
(Calidris melanotos)

Look for boldly striped brown breast sharply demarcated from white belly. Bill is short and dark. Legs yellowish or greenish. Male, with large gullar sac, makes whooping call as it flies over tundra. Nests among dense sedges or grasses on slightly raised, drier sites in wet sedge tundra or along mounds or ridges. Nest cup is a shallow depression lined with matted grasses and sedges, lichen, moss, and willow or dwarf birch leaves. Incubation by female only.

Relative Abundance: Southwest–rare, Y-K Delta–rare, Northwest–uncommon, Arctic Coastal Plain–abundant

Typical Clutch: 4 eggs
Average Egg Size: 37.4 x 26.4 mm

Max
Min

Actual Size

Buff-breasted Sandpiper
(Tryngites subruficollis)

Clean buffy breast, yellow legs, short bill. Nests on moist to dry tundra, usually in grass or sedge meadows. Nest is a shallow scrape sparsely lined with grasses or willow leaves. Eggs pale creamy, possibly tinted greenish or olive; heavily blotched and spotted in brown; markings concentrated at larger end, often with spiral elongation. Incubation by female only.

Relative Abundance: Nests only on Arctic Coastal Plain–uncommon, but abundance is variable among years

Typical Clutch: 4 eggs
Average Egg Size: 38.1 x 27 mm

Actual Size

SHOREBIRDS

Stilt Sandpiper
(Calidris himantopus)

Small shorebird with long legs, neck, and bill. Heavily barred underparts and chestnut cheek patch. Legs blackish green to olive. Call sounds like a donkey "*xxree-xxree-xxree-xxree-ee-haw, ee-haw*." Nest is a shallow hollow in sedge or moss lined mostly with willow leaves; also grasses, moss, lichens, sedges, cotton grass. Eggs pale creamy-olive or creamy-buff, blotched and spotted with reddish- or purplish-brown and pale purple. Markings usually concentrated at larger end. Incubation by both sexes.

Relative Abundance: Nests only on Arctic Coastal Plain—common

Typical Clutch: 4 eggs
Average Egg Size: 37 x 25.6 mm

Actual Size

Baird's Sandpiper
(Calidris bairdii)

Slender shorebird with thin, straight bill, dark legs; back is silvery with black spots. Wings extend beyond tail. Nests on dry, sparsely vegetated tundra. Nest is a shallow hollow sparsely lined with leaves and grass. Eggs buff, spotted or mottled with reddish-brown; more densely marked at larger end often with distinct spiral elongation in markings. Incubation by both sexes.

Relative Abundance: Southwest–none, Y-K Delta–rare, Northwest–rare, Arctic Coastal Plain–uncommon

Typical Clutch: 4 eggs
Average Egg Size: 33 x 23.8 mm

Actual Size

SHOREBIRDS

Rock Sandpiper
(Calidris ptilocnemis)

Small shorebird with black patch on lower breast (not on belly like dunlin), greenish-yellow legs, distinct dark cheek patch in male, bill slightly downcurved and black with yellowish-greenish at base. Nests commonly in heath tundra, unlike dunlin. Nest is a deep hollow lined with grasses or leaves. Eggs pale green or olive; boldly spotted and blotched with brown, usually concentrated at larger end. Incubation by both sexes.

Relative Abundance: Southwest–common, Y-K Delta–common, Northwest–rare, North Slope–none

Typical Clutch: 4 eggs
Average Egg Size: 38.6 x 27.2 mm

Max
Min

Actual Size

Dunlin
(Calidris alpina)

Small shorebird with conspicuous black belly patch. Sturdy bill droops slightly at end. Has reddish back, dark streaked breast, brown crown streaked with chestnut, black legs. Call is a harsh reedy "*kreee.*" Nests mostly in grass or sedge meadows. Nest is a cup-like hollow in a clump of grass or sedge lined with dry vegetation. Eggs pale olive or greenish; variably blotched, spotted, or speckled with dark brown. Egg markings usually concentrated at larger end. Incubation by both sexes.

Relative Abundance: Southwest–common, Y-K Delta–abundant, Northwest–common, Arctic Coastal Plain–abundant

Typical Clutch: 4 eggs
Average Egg Size: 36 x 25.9 mm

Max
Min

Actual Size

SHOREBIRDS

Western Sandpiper
(Calidris mauri)

Small sandpiper with black legs and slightly drooping bill. Has spotted sides; rufous at base of scapulars, on crown and ear patch. Similar to semipalmated sandpiper, but semipalmated lacks rufous. Call is a high raspy "*jeet.*" Nests most commonly in upland tundra. Nest is a shallow scrape/cup lined with lichen fragments or leaves, often partially hidden by overhanging dwarf birch or grass. Eggs creamy-white to buffish or brownish; heavily speckled, spotted, and blotched with reddish-brown. Markings concentrated at larger end. Incubation by both sexes.

Relative Abundance: Southwest—uncommon, Y-K Delta—abundant, Northwest—common, Arctic Coastal Plain—rare

Typical Clutch: 4 eggs
Average Egg Size: 31.4 x 21.5 mm

Max
Min

Actual Size

Semipalmated Sandpiper
(Calidris pusilla)

Small sandpiper with black legs and straight stubby bill. Similar to western sandpiper, but has slightly shorter bill, lacks spotting on flanks and has only a tinge of rufous on crown, ear patch, and scapulars. Also, feathers on its back have large black centers outlined in gray-cream; not rufous as in western. Call is a short "churt." Generally nests in grass-sedge near the coast and in uplands farther inland. Eggs pale olive, greenish, or buff, heavily spotted with reddish-brown or sparsely spotted and blotched with dark brown or purple-brown. Egg color and markings highly variable. Incubation by both sexes.

Relative Abundance: Southwest–none, Y-K Delta–abundant, Northwest–common, Arctic Coastal Plain–abundant

Typical Clutch: 4 eggs
Average Egg Size: 30 x 21.3 mm

Max
Min

Actual Size

Least Sandpiper
(Calidris minutilla)

Small shorebird with yellowish green legs and short fine black bill. Nests in open wet areas, usually on or inside a hummock or tuft of vegetation. Nest is a hollow lined with dead leaves, grasses, or stems. Eggs pale buff, spotted or blotched in reddish-brown. Markings heavier at larger end. Incubation by both sexes.

Relative Abundance: Southwest–common, Y-K Delta–rare, Northwest–none, Arctic Coastal Plain–none

Typical Clutch: 4 eggs
Average Egg Size: 29 x 21 mm

Max
Min

Actual Size

Semipalmated Plover
(Charadrius semipalmatus)

Small shorebird easily distinguished by black breast-band around neck. Legs yellow to dull orange. Nests in open sites on gravel, rocky tundra, sand, lichen, or mudflat. Nest is a shallow depression lined with whatever material is present around nest site. Often very tolerant of intruders. Incubation by both sexes.

Relative Abundance: Southwest–common, Y-K Delta–uncommon, Northwest–uncommon, Arctic Coastal Plain–rare

Typical Clutch: 4 eggs
Average Egg Size: 33.2 x 23.6 mm

Max
Min

Actual Size

SHOREBIRDS

71

Red Phalarope
(*Phalaropus fulicaria*)

Bill is yellow, shorter and thicker than red-necked phalarope's. Bold white wing stripe shows in flight. Has white face patch and red body; males are less colorful than females. Call is a whistling "*wit*," higher and sharper than red-necked's; also a "*breeep!*" Nests in sedge meadows, often along shorelines of ponds. Nest is a cup-like hollow in a clump of sedge or grass lined with grass. Eggs light olive or green with irregular blotches, spots, and specks of black and blackish-brown; similar to red-necked phalarope. Usually has a few large blotches combined with profuse tiny markings. Incubation by male only.

Relative Abundance: Southwest–none, Y-K Delta–common, Northwest–uncommon, Arctic Coastal Plain–abundant

Typical Clutch: 4 eggs
Average Egg Size: 32 x 23 mm

Actual Size

Max
Min

Red-necked Phalarope
(Phalaropus lobatus)

Bill is slender. White wing stripe shows in flight. Males are less colorful than females. Call is a high sharp "*kit.*" Nests in sedge meadows often along shorelines of ponds. Nest is a cup-like hollow in a clump of sedge or grass, lined with grass. Eggs light olive or green with irregular blotches, spots, and specks of black and blackish-brown. Usually has a few large blotches combined with profuse tiny markings. Incubation by male only.

Relative Abundance: Southwest–common, Y-K Delta–abundant, Northwest–common, Arctic Coastal Plain–common

MALE GMB

Typical Clutch: 4 eggs
Average Egg Size: 29.6 x 20.8 mm

Max
Min   Actual Size

SHOREBIRDS

Redpolls

Redpoll taxonomy is not settled. Adults of **hoary redpoll** *(Carduelis hornemanni)* appear similar to **common redpoll** *(Carduelis flammea)* and both species are highly variable; eggs and nests are practically identical. Adult has conical pointed bill, black throat, streaked flanks. Nests in crotches of shrubs, usually willow, and locally on man-made structures that provide nest support. Nest is built of coarse grasses and weed stems, cotton from cotton grass or willow; lined with ptarmigan feathers or fur. Eggs pale greenish to pale blue, spotted with purple, particularly at larger end. Incubation by female only.

Relative Abundance: Southwest–uncommon, Y-K Delta–common, Northwest–common, Arctic Coastal Plain–uncommon

Typical Clutch: 4-5 eggs
Average Egg Size: 18.2 x 13 mm

Actual Size   Max / Min

Snow Bunting
(Plectrophenax nivalis)

FEMALE **MALE**

Both males and females show flashy white wing patch, black tail feathers with outer white feathers, white underparts. Nests underneath things or in cavities, particularly in rocky areas; usually difficult to find. Thick-walled nest of moss and grasses, lined with fine grasses and often fur and white feathers. Eggs pale blue or greenish-blue, variably blotched, spotted, or speckled with various shades of brown. **McKay's bunting** *(Plectrophenax hyperboreus)* is similar, but nests only on St. Matthew and Hall islands in Bering Sea. Incubation by female only.

McKay's bunting—FEMALE FWS

Relative Abundance: Uncommon in all areas

nest site

GWB

FWS

Typical Clutch: 4-6 eggs
Average Egg Size: 22.9 x 16.2 mm

Max
Min

Actual Size

PASSERINES

Lapland Longspur
(Calcarius lapponicus)

FEMALE MALE

DRR

TD

EH

Female has dark cheek patch and dark streaks on upper breast forming a band. Outer tail feathers are partly white. In presence of human intruder, male makes a wheezy "*dzeeu*" alarm call, whereas female makes a high-pitched "*si-si-si-si.*" Nests on open ground or sometimes in the side of a mound. Nest is a cup-like hollow in a clump of grass or sedge lined with fine grass, hair, or feathers. Eggs pale greenish, buffish, or grayish, smooth and glossy with indistinct mottling of dull reddish-brown or purplish-brown. Incubation by female only.

Relative Abundance: Southwest–common, Y-K Delta–abundant, Northwest–common, Arctic Coastal Plain–abundant

Typical Clutch: 5-6 eggs
Average Egg Size: 21 x 15 mm

Actual Size

Max
Min

Savannah Sparrow
(Passerculus sandwichensis)

Adult has yellow in front of eye, finely streaked breast, and white belly. Song begins with two or three "*chip*" notes followed by two buzzy trills, the second trill lower and briefer. Typically nests on ground in grass meadows; sometimes found along shorelines or slough banks. Nest is a cup-like hollow lined with fine grass, hair, or feathers; concealed by overhanging vegetation, usually grasses. Incubation by female only.

Relative Abundance: Southwest–common, Y-K Delta–abundant, Northwest–common, Arctic Coastal Plain–common

DRR

Typical Clutch: 4-5 eggs
Average Egg Size: 19 x 15 mm

Max
Min

Actual Size

Golden-crowned Sparrow
(Zonotrichia atricapilla)

Fairly large, long-tailed sparrow with a black crown and forehead divided by a yellow stripe, a bicolored bill, and a dull gray breast. Nests in shrubby-tundra habitats, mostly on ground. Nest is a thick cup of twigs and coarse grasses lined with fine grass and sometimes hair or ptarmigan feathers. Egg color is variable, but usually pale blue or greenish blue with fine reddish-brown markings. Incubation by female only.

Relative Abundance: Southwest–common, Y-K Delta–uncommon, Northwest–uncommon, Arctic Coastal Plain–none

Typical Clutch: 3-5 eggs
Average Egg Size: 23.1 x 16.5 mm

Actual Size   Max / Min

Yellow Wagtail
(Motacilla flava)

Look for yellowish breast, wagging tail, and white outer tail feathers. Call is a buzzy "*tzeeu tzeeu tzeek.*" Nests more frequently inland from coast, on ground in shrubby areas, usually willow, typically at edge of grassy areas and rivers/sloughs. Nests often placed under mats of overhanging vegetation in dirt banks. Nest is a cup composed mainly of grasses, often lined with hair or ptarmigan feathers. Eggs pale buff or grayish; heavily, finely, and uniformly speckled with yellowish-buff. Incubation by both sexes.

Relative Abundance: Southwest–uncommon, Y-K Delta–uncommon, Northwest–uncommon, Arctic Coastal Plain–rare

Typical Clutch: 4-6 eggs
Average Egg Size: 18.8 x 15 mm

PASSERINES

Sizing Chart for Unpatterned (Plain) Eggs

Egg sizing charts provide a quick reference to help reduce the number of prospective species to which an egg could belong. There is considerable variation in egg size both within and among species, so this provides only an approximate guide. Refer to species pages for additional information on egg and nest characteristics.

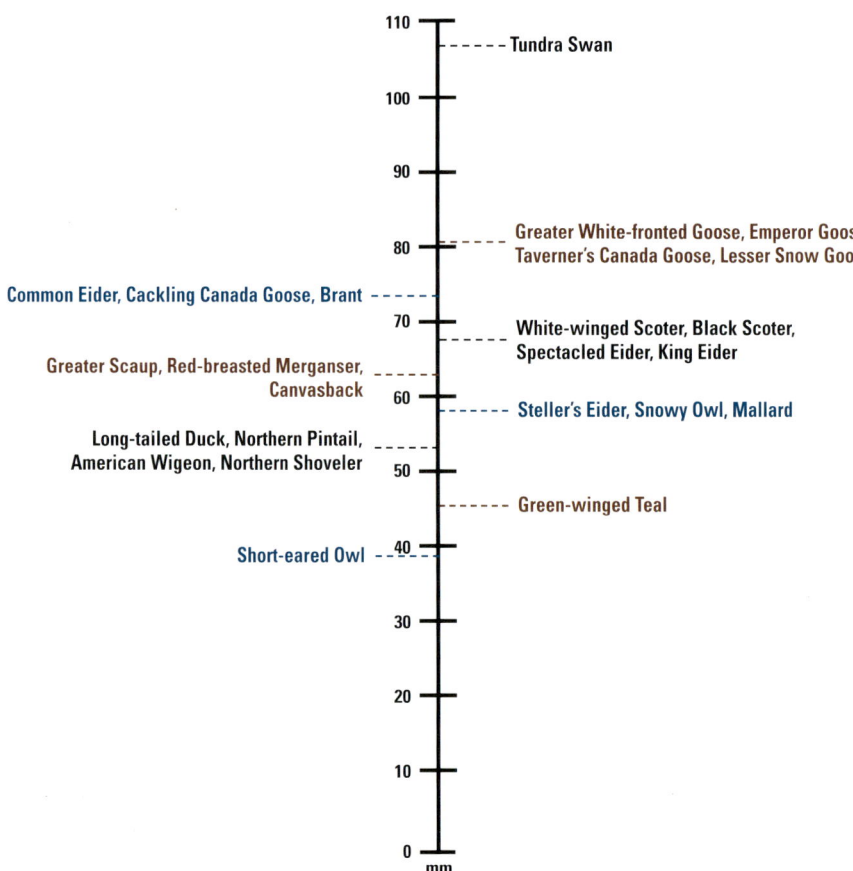